they are generally transferred from organism to organism in food-chains. One of the early examples was at Clear Lake in the United States where TDE, a pesticide similar to DDT, had been applied to control the aquatic gnat *Chaoborus*. Grebes started dying five years after the first TDE application and were found to contain up to 1·6 μg TDE/g of visceral fat (organochlorine pesticides are fat-soluble). The TDE was taken up from fish on which the birds were feeding.

Man contains traces of several organochlorine pesticides and average concentrations of 26, 12, and 3 μg DDT/g of body fat have been found in India, the United States, and the United Kingdom respectively. These are now declining in many countries and seem to be harmless. Workers in the pesticide industry contain much higher concentrations and do not suffer adverse effects.

Some toxic materials with cumulative effects may remain undetected for decades. The poly-chlorinated diphenyls were used extensively by the paint and adhesives industries for over thirty years before it was suggested that they contributed to widespread mortalities such as that of sea-birds off the west coast of Britain in 1969. Mercury has been widely used not only as a fungicide but also as a catalyst for various industrial processes for many years and only now are we becoming aware of the high levels that are accumulating in some animals, particularly fish in certain coastal areas.

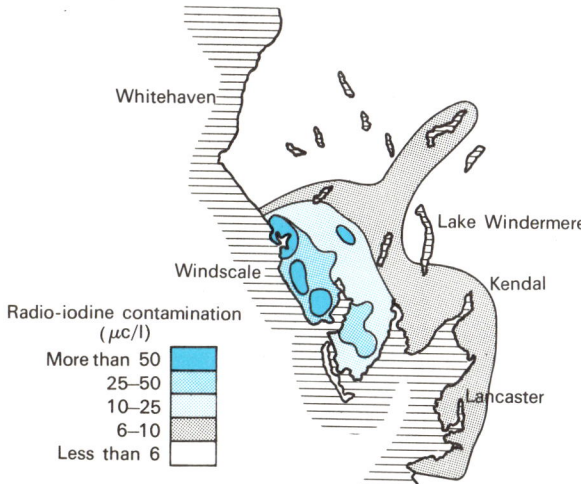

Whitehaven
Lake Windermere
Windscale
Kendal
Radio-iodine contamination
(μc/l)
More than 50
25–50
10–25
6–10
Less than 6
Lancaster

FIG. 1. Map showing contours of radio-iodine contamination of milk three days after a reactor accident at Windscale in 1957 when fission products were released to the atmosphere. (From *The hazards to man of nuclear and allied radiation*, Cmnd 1225; by kind permission of the Controller H.M.S.O.)

AIR POLLUTION

Sources

Air pollution results largely from the products of combustion and of materials carried with the hot gases derived from the burning of fuels. Most of man's current energy requirements are derived from fossil fuels: coal, oil, and natural gas. In the future an increasing proportion will be derived from nuclear sources but because of the rapid increase in demands for energy it does not seem likely that the use of fossil fuels will decline over the next few decades.

At present nuclear energy is obtained by splitting atoms (fission) which results in the aerial discharge of radioactive gases, principally krypton-85 (half-life 10 years) and tritium (half-life 12 years). These are at present unimportant but the escalation of nuclear power will necessitate their more effective removal and disposal. Accidents have occurred at reactors. At Windscale, Cumberland, in 1957 isotopes of iodine, tellurium, caesium, and strontium were released into the atmosphere: most deposits fell in the neighbouring areas but low concentrations of radioactivity were detected as far away as Germany and Norway. Fig. 1 shows the extent of radio-iodine contamination in milk collected near Windscale, much of which was thrown away. Iodine is probably the most hazardous radioactive contaminant for it is concentrated by the thyroid gland. Strontium is concentrated in the skeleton, replacing calcium in developing bone tissues. There are hopes that nuclear fusion (the basis of the H-bomb) will eventually be harnessed as a power supply and it is a much cleaner process than nuclear fission.

Fossil fuels when combusted produce a range of carbon, nitrogen, and sulphur compounds, the precise composition of the resulting gases depending mainly on the degree of oxidation and the composition of the fuel. In domestic and industrial burners the principal pollutants are carbon dioxide, smoke, and oxides of sulphur and nitrogen. It is likely that between A.D. 1890 and 2000 there will have been an 18-fold increase in the rate of emission of carbon dioxide in the United States resulting from the combustion of fossil fuels. Similar increases of sulphur and nitrogen compounds are likely to occur unless steps are taken to remove them during fuel preparation or combustion. Emission of industrial smoke in Great Britain has been greatly reduced by improvements in the design of industrial combustion units, stimulated

3

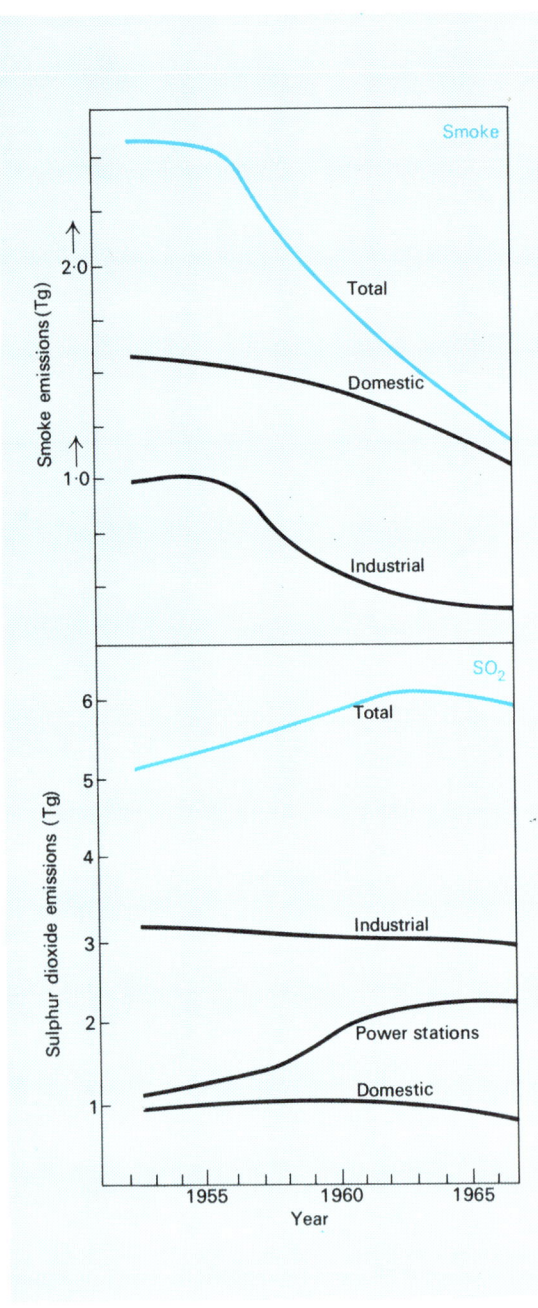

Fig. 2. Changes in emissions of smoke and SO₂ in Britain between 1951 and 1967. (From Craxford and Weatherley (1968). *XXI Int. Congr. CEBEDEAU.*)

largely by anti-pollution legislation; the biggest contribution now comes from domestic sources (Fig. 2). SO_2 emission continued to increase until the mid-1960s (fossil fuels generally contain between 1 and 4 per cent sulphur) but has stayed fairly steady since (Fig. 2). SO_2 concentration at ground level has declined recently in the more polluted urban areas as a result of the installation of higher industrial chimneys which help to disperse gases before they reach ground level. SO_2 can be removed from flue gases using limestone or dolomite, but there is a substantial difference between the value of sulphur, less than £10 per tonne, and the cost of removal, at least £25 per tonne. Sulphur can also be removed from fuels by reducing it to hydrogen-sulphide and this is somewhat less costly.

The exhaust gases from motor vehicles contain, as a result of incomplete combustion, hydrocarbons, carbon monoxide, and various oxides of nitrogen and sulphur. Vehicles probably contribute more than 80 per cent of the global emission of CO and in busy streets a concentration of 70 μg/m³ is not uncommon. Lead which is added to most motor fuels is also present in fairly high quantities. In 1967 about 3000 tonnes of tetramethyl- and tetraethyl-lead was produced in the United States alone and of the 0·5 g lead/litre added to petrol about 60 per cent was emitted in exhaust. Improvements in the design of carburation and exhaust systems can reduce the discharge of CO and hydrocarbons to about 40 per cent of the 1962 levels at small cost, but additional improvements (reduction of nitrogen oxides, and lead) pose complex design problems.

Dispersion

The dispersion of air pollutants depends on meteorological conditions, the conditions of emission (height above ground, temperature, etc.), the chemical stability of the pollutants, and on the other constituents of the air.

Large heavy particles carried up chimneys with hot gases soon settle and about one hundred times as much deposit is collected in heavily industrialized areas as in neighbouring areas of countryside. In one recent investigation it was found that oak leaves about 200 m from an industrial site had 200 μg deposit/cm² of leaf compared with 40 μg/cm² only 400 m away from the site. Whilst much of the lead emitted from vehicles remains quite near the road-sides, contaminating ground and vegetation, smaller particles travel much greater distances and may contaminate appreciable areas. Samples taken in the Pacific Ocean and the Greenland Ice-cap show the widespread presence of lead in the atmosphere; concentrations in such isolated areas have increased 30-fold since pre-industrial times.

The highest concentrations of pollution are generally experienced near the source during an inversion, when the air near ground level is denser, because it is cooler, than the air above it. This frequently occurs at night, when heat losses from the ground exceed incoming radiation, particularly in sheltered valleys or low-lying areas. Pollutants become trapped in the cool stable ground layer unless emitted from tall chimneys which protrude through it or unless the plume, because of its high temperature, is carried above the inversion. Sometimes pollutants collect at the ceiling of the inversion layer and when ground temperature rises thermal conduction produces turbulence. High concentrations of pollutants are then brought down to ground level (fumigation). Smog is generally associated with inversions and occurs when polluted air containing particulate matter is cooled below the dew point.

If pollutants are short-lived, becoming rapidly changed to innocuous materials, there is only need to consider their distribution near the source. Most gaseous pollutants however have a life of at least several hours (e.g. SO_2) and some even years (e.g. CO), and being diluted only by dispersion, can affect large areas. SO_2 is initially oxidized to SO_3, this being accelerated by sunlight, NO_2, certain hydrocarbons, and ions of iron and manganese. SO_3 reacts with water to produce sulphuric acid and with ammonia to form ammonium sulphate. The increasing acidity of rain water in Holland and of some lakes in Sweden may be due in part to the SO_2 emitted in Britain, though the increase in atmospheric SO_2 over the past ten to twenty years could only account for 0·1 of the noted 1·0 change in pH.

Interaction of air pollutants produces the 'photochemical smog' of Los Angeles, Chicago, and Tokyo. NO_2 absorbs ultraviolet radiation from the sun and in its breakdown to NO, ozone is formed which reacts with certain hydrocarbons to form secondary pollutants like peroxyacyl nitrates.

Effects

Effects on the atmosphere. Weather conditions are substantially affected by air pollution. The formation and persistence of fog depend on the density of suspended particles which act as nuclei on which droplets form. The incidence of fogs in London has declined markedly since the Clean Air Act of 1956.

It is not known to what extent pollution affects rainfall though lead particles discharged in vehicle exhaust gases, when exposed to the minute quantities of iodine vapour that are present in the atmosphere, produce large numbers of the freezing nuclei which form the basis of cloud-seeding operations for the promotion of rain.

Polluted air generally reduces the amount of short-wave solar radiation, visible and ultraviolet, which reaches the earth's surface. The loss is strongly dependent on wavelength, the shorter lengths being most seriously affected. During one smog in Pasadena, California, shorter wavelength intensities were reduced by at least 90 per cent. This was attributed to scattering by particles and to absorption by ozone and NO_2.

On a global scale it has been suggested that pollution is affecting the heat balance of the earth but two types of pollutants seem to be antagonistic in this respect. Atmospheric particles reflect long-wave radiation (heat), reducing the amount that reaches the earth's surface; but the 15 per cent increase in CO_2 concentration between 1900 and 1970 reduces the reflection of heat from the earth.

There has recently been a decline in the surface temperature of the earth, and this may be attributable to a rise in concentration of atmospheric particles by pollution or by volcanic eruption or by soil erosion.

Effects on plants. A global estimate of the pollution damage to crops has not been made but in California, which suffers particularly from photochemical smog, losses from direct injury and impairment of growth and quality probably exceed £50 million per year. It is often difficult to separate the effects of different pollutants, and ozone and SO_2 (for example) when individually present in amounts too low to cause damage, have a marked effect on the tobacco crop when present together.

Sulphur dioxide. A study in Leeds showed that the growth of radish and lettuce planted in a standard soil was greatly reduced in the more polluted part of the city. Later studies in London with several garden flowers and in Manchester with rye-grass showed similar reductions in yield and, in some cases, tissue damage. In the Manchester study the reduction in yield was attributed principally to SO_2 at concentrations as low as 60 µg/m³—a

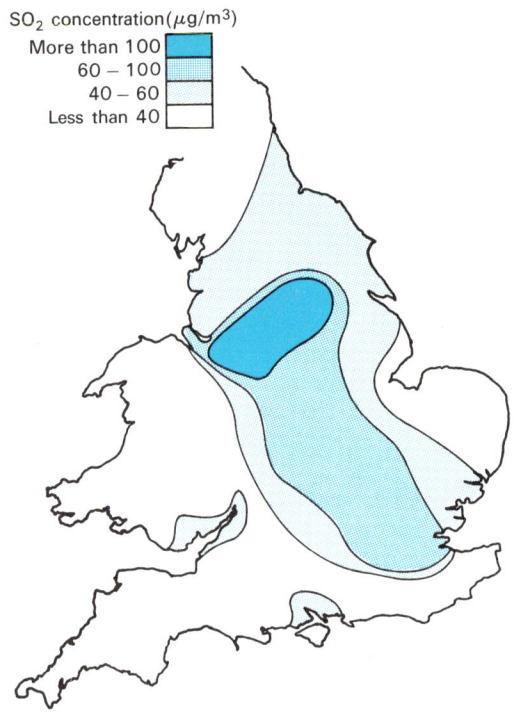

SO$_2$ concentration(μg/m^3)
More than 100
60 – 100
40 – 60
Less than 40

FIG. 3. Average winter concentrations of SO$_2$ in rural areas of England and Wales, 1968–9. Although there is much variation in SO$_2$ concentrations, particularly near towns, this map shows the general levels of pollution over broad areas. (Data from *National survey, smoke and sulphur dioxide* 1968–9.)

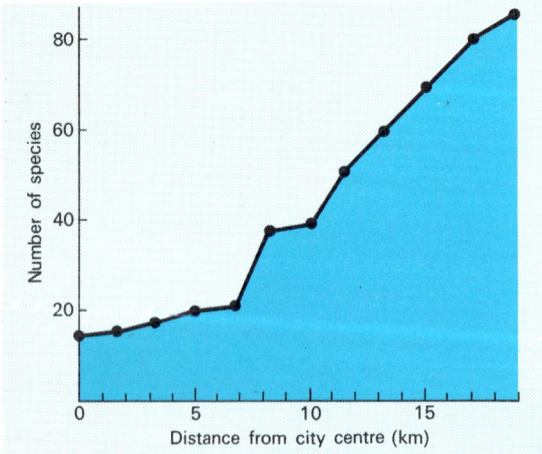

FIG. 4. The number of species of lichen and moss increases with distance to the west from the centre of Newcastle. (From Gilbert (1968), *I. Europ. Conf. Influence Air Pollut. Plants Animals.*)

concentration found over very large areas of Britain, particularly in the winter (Fig. 3).

At least one dubious advantage of SO$_2$ pollution has been reported. Blackspot of roses (*Diplocarpon rosae*) is severely checked or even eliminated in areas where SO$_2$ concentrations exceed 100 μg/m^3.

Perhaps the most dramatic vegetation damage is associated with the smelting of ores. This gives off both SO$_2$ and significant amounts of metals. In several parts of the United States extensive vegetation deserts have been produced. The destruction of plants and the acidic nature of the rain has led to a breakdown and erosion of the soil. At Ducktown, Tennessee, the emission of pollution was drastically reduced in 1905 but 50 years later the damage was still extensive despite attempts to replant the area. In a copper-smelting area of Sacramento Valley, California, all vegetation was killed over an area of 260 km^2 and growth was severely affected over a further 320 km^2. A study of another copper-smelting region in Ontario showed adverse effects on vegetation extending for about 32 km from the smelter although pronounced accumulation of sulphate only occurred within the soil and soil drainage water for a distance of about 3 km. The most tolerant species included oak (*Quercus rubra*), maple (*Acer rubrum*), and elder (*Sambucus pubeus*); the most sensitive were a pine (*Pinus strobus*) and bilberry (*Vaccinium myrtilloides*).

Bryophytes and lichens absorb and accumulate many substances in their thalli or leaves, and are particularly sensitive to air pollution, especially SO$_2$. Gilbert found in north-east England a bryophyte and lichen desert over an area of 1000 km^2, centred on the Newcastle coalfield. The number of species declined steadily nearer Newcastle (Fig. 4), though a few lichens (e.g. *Lecanora conizaeoides*, *L. dispersa*) and mosses (e.g. *Ceratodon purpureus*, *Bryum argentum*, *Funaria hygrometrica*) tolerated conditions in the centre of the city. The lichen *Grimmia pulvinata* and the moss *Hypnum cupressiforme* were not found on sandstone walls above an annual average SO$_2$ concentration of 45 μg/m^3, nor the lichens *Parmelia saxatilis* and *P. fulginosa* above 60 μg/m^3. A pollution map of England and Wales based on a wide variety of such indicators has now been prepared, and will enable landscape designers to select appropriate trees for urban and industrial areas.

Sulphur dioxide is converted in most plants to sulphate. Generally this is relatively non-toxic,

6

provided that the rate of SO_2 input is not too high. If input exceeds conversion SO_2 remains as the far more toxic sulphurous acid and sulphite, and it appears that photosynthesis is disrupted. This suggests that the peak concentrations of SO_2 experienced by plants are far more critical than the average concentrations which are determined in most monitoring programmes.

Particles. Fine particles reduce the amount of light available for photosynthesis when in suspension and when covering the leaf surface. Settled material blocks stomata and so interferes with the normal processes of water and gas exchange. A recent study of the effects of pollution from a smokeless-fuel plant on a neighbouring oak-wood showed that up to 60 per cent of the stomata of oak leaves were completely or partially blocked during the summer (Fig. 5). The blockage was related to the amount of deposit covering leaves and to the distance from the polluting source. Other changes in leaf anatomy included reduced size, thickened cuticle, and a more closely packed palisade.

Smog. Ozone, peroxyacyl compounds such as peroxyacyl nitrate, and nitrogen oxides occur in photochemical smog and damage plants. In the United States they have caused tissue damage to lettuce and spinach and substantial reductions in growth and yield of tomatoes, citrus trees, tobacco, and conifers. Ozone attacks the palisade cells which lose water and disintegrate. Peroxyacyl nitrate generally attacks cells of the spongy mesophyll, particularly those adjacent to the lower epidermis. Although early work showed that NO_2 caused damage only at concentrations above 5.7 mg/m^3, recent studies indicate that much lower concentrations over periods longer than a few hours can depress growth. Atmospheric NO_2 concentration rarely exceeds 2.9 mg/m^3.

Fluorides. Fluorides are emitted in quantity by only a few industries (aluminium, steel, brick-making, phosphate fertilizer). They can have a local damaging effect on some plants, particularly young conifers. Fluorides are emitted as gas (HF and SiF_4) and dust, the gas being far more damaging. Fluoride is a cumulative poison which enters the leaf through stomata and moves in the water stream to the leaf margin where it is concentrated. Injury seems partly due to the inhibition of enzymes concerned with cellulose synthesis. Some plants, including many grasses, are relatively resistant to fluorides and near an aluminium factory in Scotland no visible damage to vegetation was

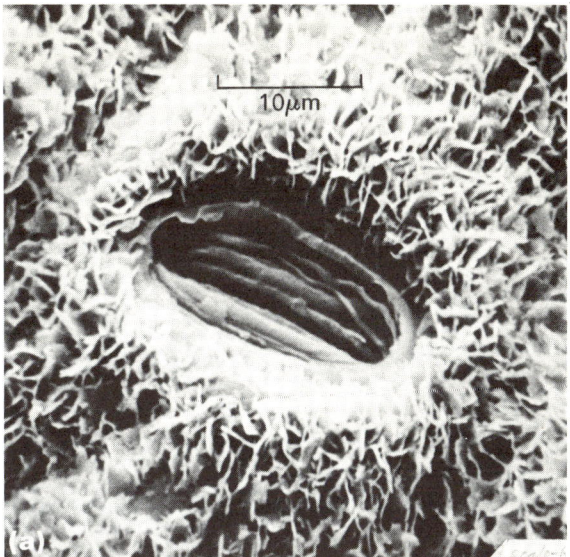

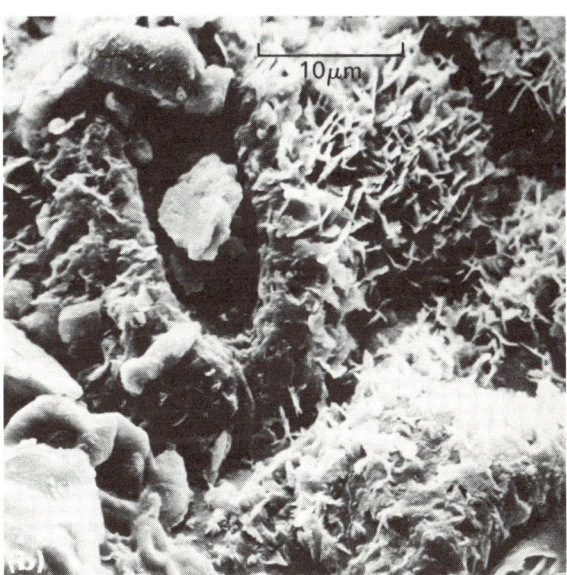

FIG. 5. Leaf surfaces seen with the scanning electron microscope. (a) Sessile oak (*Quercus*) with stoma, from an unpolluted area. (b) The same, from an area with high fall-out of solids. The fall-out debris clearly interferes with the opening and closing of stomata. (Long Ashton Research Station, Bristol.)

found when its fluorine content was as high as 1 mg F/g dry weight of plant.

Effects on man and other vertebrates. The average man requires about 13 kg of air per day, and as might be expected the primary target organ for most air pollutants is the lung, which presents the only extensive surface area of poorly protected tissue.

Experimental studies with animals have revealed much about mechanisms of pollutant action and have given a rough guide to the likely effects of different patterns of pollution on man. Of the pollutants present in air only CO and perhaps SO_2 may occasionally reach levels that are acutely dangerous. What is perhaps of more concern, and here laboratory experiments become difficult, are the chronic effects of exposure to rather lower concentrations of several pollutants acting together.

Dramatic disasters are associated with smog, when high concentrations of several pollutants persist for hours or even days. The London smog of December 1952 lasted five days during which the smoke and SO_2 concentrations reached 4·4 and 3·7 mg/m^3 in one area and about 4000 people died in excess of those expected at that time of year. Mortality from SO_2 is principally due to irritation of the respiratory system in people suffering from chronic diseases of the heart and lungs.

Mortality statistics also suggest that other deaths are linked with air pollution. Chronic bronchitis (the English disease) is prevalent in the damp foggy climate of Britain causing a death rate in England and Wales of 55 per 100 000. This compares unfavourably with 2 per 100 000 in the United States and 4 per 100 000 in Denmark. Regional studies in Britain show that atmospheric pollution is one major cause.

Whilst lung cancer is exacerbated by cigarette smoking there is also an 'urban factor' which although complex is probably associated, at least in part, with air pollution. Carcinogens such as benz-pyrenes have been detected in polluted air and in their presence inert particulate material may stimulate cancer growth.

The short-term effects of photochemical smog seem to be caused mainly by ozone and are limited to discomfort and irritation (headaches, dryness of throat). Ozone concentrations rarely exceed 0·3 mg/m^3, which is around the limit of biological detection. At the highest ozone concentration yet recorded, 2 mg/m^3, there are no serious effects on man, but above this level pulmonary congestion can occur and the dissociation of oxyhaemoglobin is impeded. Some animals appear to be more sensitive to ozone than man: mice exposed to only 0·4 mg/m^3 developed some destruction of heart muscle and had fewer and smaller litters; young chicks died after continuous exposure to 2 mg/m^3 for 5 days. Studies with rats at higher concentrations have revealed RNA and DNA changes in lung and liver tissue.

As with plants, the effects of the highest observed concentrations of NO_2 (2 mg/m^3) are not harmful. Mortalities in mice have been produced only by continuous exposure for 3 months to concentrations of about 9 mg/m^3.

The SO_2 in smog is said to be dangerous because it stimulates coughing in those suffering from heart and lung diseases, but pure SO_2 even at the high concentration of 3·75 mg/m^3 has little effect on coughing. In polluted air SO_2 is generally present with other pollutants and animal studies have shown that sulphuric acid mist and particles of salt (NaCl) greatly enhance SO_2 effects, particularly when the droplets and particles are very small. This enhanced action or *synergism* of mixtures of pollutants probably accounts for the severe effects of smog.

Carbon monoxide is the atmospheric pollutant giving rise to greatest concern. It is frequently found at concentrations above 20 mg/m^3, and 400 mg/m^3 was once recorded in London. In the blood it forms carboxyhaemoglobin, so reducing its oxygen-carrying capacity. Ten per cent carboxyhaemoglobin in the blood of a healthy person is the physiological limit beyond which damage may be caused, but many people have impairment from such other causes as cigarette smoking, anaemia, and lung disease, and the safe limit for carbon monoxide in the atmosphere has been set at 33 mg/m^3 which is equivalent to 5 per cent carboxyhaemoglobin in the blood. Blood takes many hours to equilibrate with this atmospheric concentration of CO and higher concentrations may be tolerated for short periods. The chronic effects of carbon monoxide are apparently negligible: carboxyhaemoglobin dissociates when CO concentrations in the air decline and the body can deal with low concentrations of CO by gradually increasing the oxygen-carrying capacity of the blood.

Although lead is being used in appreciable amounts in fuels there is no evidence that the concentration in urban air is causing harmful effects. The major human intakes of lead are in food and water, probably amounting to about 0·1 to 0·3 mg/person/day. Air generally contributes between 0·01 and 0·1 mg/person/day. However, much higher proportions of lead are absorbed through the lung than the stomach and intestine and the blood lead levels are generally higher in urban than rural areas. There is some evidence that lead ingested in vegetation has been responsible for the deaths of horses in the Swansea Valley, South Wales, although the source of the lead has not been established. It is possible that neighbouring reclamation schemes, in which tips containing lead have been moved, have resulted in the dispersion of fine particles throughout the area.

Fluorine poisoning (fluorosis) is not a problem except to herbivorous animals in regions near concentrated sources where vegetation is heavily contaminated either by fall-out of particles or by absorption of gases containing fluorine. Damage to teeth and bones prevents grazing and so reduces meat production and milk yield. In severely affected areas vegetation may contain up to 0·2 per cent fluorine and animals up to 2 per cent in bone.

FIG. 6. Simplified flow-diagram of a sewage-works. Sewage, after passing through a screen or comminutor (A) to break up debris, flows into a primary settlement tank (B) where many solids settle, then into oxidation units (C), and finally into further settlement tanks (D) before being discharged to a river or estuary. In many modern sewage-works additional treatment is given using filters (E) to remove the particulate material which does not settle readily.

WATER POLLUTION
Domestic use: sewage

In Britain each person uses domestically about 140 l water/day: in the United States current consumption is about 230 l/day. Except in some rural areas this domestically-used water, sometimes mixed with industrial waste-waters, is transferred to sewage-works for treatment. Sewage-works are designed principally to remove solid material by settlement and to oxidize organic matter to carbon dioxide and oxides of nitrogen and sulphur. Fig. 6 shows a simplified flow-diagram of a sewage-works. Two systems of oxidation, activated sludge and percolating filters, are commonly used and both depend on micro-organisms (Fig. 7). Neither is completely efficient for whilst most of the organic matter is oxidized some is built up into microbial cells which must subsequently be removed, generally by additional settlement tanks. In the activated sludge system the liquid organic waste, inoculated by micro-organisms, flows into tanks through which air is blown to aid mixing and to provide oxygen. In percolating filters the organic waste is sprayed on to a bed of stones or corrugated plastic to which micro-organisms become attached and on which they grow. The liquid drains through the filter, where the breakdown of organic matter occurs, and passes to tanks where excess micro-organisms, mainly bacteria and fungi that have

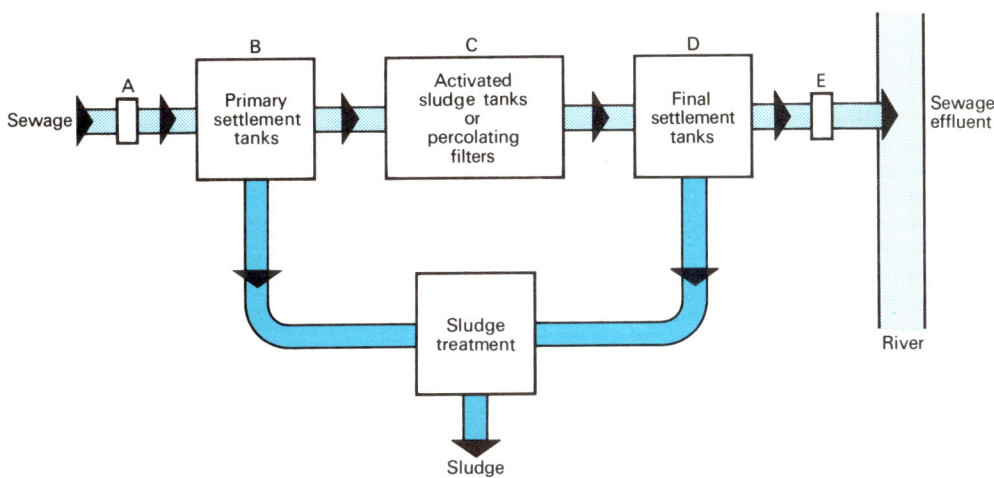

Photos by Water Pollution Research Laboratory

FIG. 7. (a) Activated sludge tanks and (b) percolating filter at sewage-works.

been removed from the filter, are able to settle. Whilst much of the carbon in many organic compounds is readily oxidized to CO_2, the oxidation of nitrogen, which is initially in the form of ammonia, is much slower. If there are insufficient microorganisms or if the temperature or retention time of sewage in the oxidation unit is low, oxidation of ammonia to nitrate does not occur and ammonia is discharged in the effluent.

The sludge collecting in settlement tanks may be treated in several ways. In some sewage-works it is fermented in the absence of air and the gaseous products like methane are burned to produce power. In others the water is squeezed out of the sludge under pressure or sucked out of it by vacuum. The dried sludge is sometimes combusted, which adds to the air pollution problem! Some sludge is disposed of profitably on agricultural land, but if it contains contaminants such as zinc and copper in too high a concentration for land application its disposal is a real problem.

In relation to their primary functions, removal of solids and organic matter, sewage-works are frequently very efficient: solids are commonly reduced from 300–400 mg/l to 10 mg/l and organic matter (expressed as organic carbon) from about 250 mg/l to 20 mg/l. Much of the organic matter which remains

in sewage effluents is very resistant to oxidation.

Organochlorine pesticides are also removed. At one sewage-works the concentration of DDT was reduced from 7 µg/l in the sewage to 0·08 µg/l in the sewage effluent—about 99 per cent removal. Sewage-works are not so efficient at removing some other potential pollutants like phosphorus and metals such as zinc and copper, and although up to 50 per cent may be absorbed by the sludge the remainder is discharged in the effluent.

There are some 1 million faecal coliform bacteria (*Escherichia coli*) per cm^3 in sewage and the usual removal rate is 90–95 per cent. Some pathogenic bacteria like that producing paratyphoid fever (*Salmonella paratyphi*) survive sewage treatment in small numbers and need subsequent treatment at the waterworks where efficient settlement and sterilization with chlorine, or occasionally ozone, removes or kills them. Protozoans, such as *Entamoeba histolytica*, which causes dysentery, and eggs of worm parasites, such as *Taenia* and *Ascaris*, have also been found in sewage-works effluents although most are removed during treatment, particularly when there is a final filtration process. Eggs of parasitic worms can survive for long periods in air-dried sludge, which is therefore a potential source of infection to grazing animals if used as a fertilizer.

Industrial use

Industrial consumption of water is almost as large as domestic and in Britain is probably equivalent to about 90 l/person/day. Some industries use very high quantities, for instance 1 tonne of steel requires between 140 and 280 kl, 1 tonne of wood pulp about 280 kl, and 1 tonne of rayon about 1000 kl. Most can be re-used by recirculation.

The quality of industrial effluents varies very much with the industry. Electro-plating wastes may contain copper, cadmium, nickel, chromate, and cyanide, all of which are very toxic and likely to kill micro-organisms in sewage-works if discharged to sewers or to kill fish and other aquatic organisms if discharged to rivers. Clearly industries having such toxic wastes must treat them before release. Other industries, for example breweries and food-processing factories, produce non-toxic wastes, mainly organic matter, that can be dealt with at sewage-works. Most pollution incidents occur when industrial toxic wastes are released 'accidentally' directly into natural waters.

Power-stations use a lot of water for cooling. At inland sites, much of the water is recirculated and heat is lost by water-evaporation to the atmosphere. Even during hot summers, river temperatures downstream of power stations in Britain rarely exceed 25°C, perhaps 5°C above normal. On estuaries and the coast, where more water is available, waste heat is discharged with the cooling water, but such high volumes are used that temperature increases even in estuaries are at the most a few degrees. This speeds up the rate of development of many organisms which is an advantage with oysters and young stages of inshore commercial fishes but a disadvantage with borers which invade wooden structures in harbours and barnacles which grow on the hulls of ships and reduce their efficiency.

Agricultural use

Irrigational needs are very variable, being principally in the summer months and in the drier regions. In England and Wales almost 800 km^2 are irrigated and in south-east England where demands are greatest, irrigation represents about 20 per cent of the total water demand of the area during the summer. Natural drainage following rain, particularly after the application of inorganic fertilizers, contains appreciable amounts of nitrate and much smaller amounts of other nutrients such as phosphorus which may cause nuisance in reservoirs and lakes by promoting the growth of algae. In a study of nutrient sources to the Great Ouse river in England it was found that over 80 per cent of the nitrogen and less than 30 per cent of the phosphorus in river water was derived from land-drainage. More phosphorus came from sewage-works, and about half of this originated in detergents.

The recent development of factory farming has created problems in the disposal of animal waste. This is about 20 times as concentrated, in terms of organic matter, as sewage. One cow produces about 10 times as much waste per day as a man, and a pig about twice as much. In one incident in Holland the whole trout population of a river was killed following the discharge of farm wastes. Treatment processes similar to those for sewage have now been developed for some of the bigger animal units.

Farmers also use considerable quantities of pesticides. In Britain in 1964 about 260 tonnes of DDT, 200 tonnes of dieldrin, and 100 tonnes of BHC were used. Generally they remain in the soil where they degrade slowly, but some get into watercourses, by the accidental dumping of pesticide containers, or in drainage waters from treated land, or in liquid disposed of after sheep-dipping. At Chew Valley reservoir, which supplies water to Bristol, the concentration of algae (*Monodus*) in the reservoir recently became so great that the filters treating the water were blocked. Disposal of residues of BHC and dieldrin from sheep-dipping near an underground watercourse draining to the reservoir had contaminated it and killed off the zooplankton that normally grazed on the alga. Although pesticides are popularly associated with agriculture by far the highest concentrations in rivers have been found in industrial areas and have come from factories where pesticides are made or where fabrics and carpets are moth-proofed.

Effects

Effects of pollution on man's drinking water. Water pollution restricts the sources that man can draw upon for supply. Quality standards for domestic use have been recommended by the World Health Organization. Lead may be toxic when the recommended concentration is exceeded. Nitrate only causes trouble in young babies; it is reduced in the gut to nitrite which is absorbed and attached to haemoglobin, preventing oxygen transfer (methaemoglobinaemia). As babies grow their stomach acidity increases and nitrate reduction is prevented. Copper, chloride, and phenol produce objection-

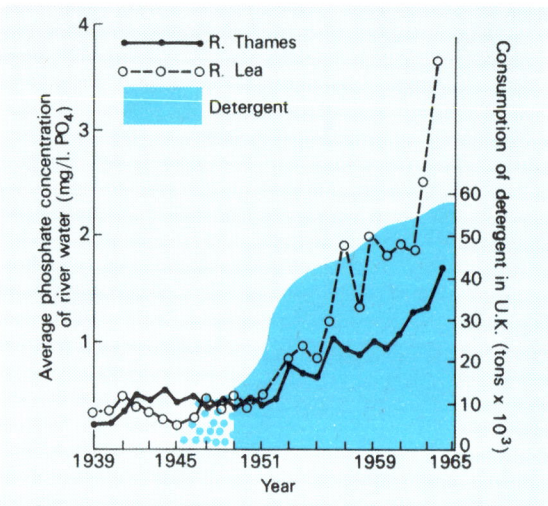

FIG. 8. Increase in phosphate concentration of two rivers after widespread use of domestic detergents. Before 1950 detergents were only used for industrial purposes. (From Owens (1968), *Wat. Res.* **2**.)

able tastes and ammonia combines with the chlorine that is used as a germicide in waterworks.

Effects on ecological balance in reservoirs and lakes. In the past, most domestic water in Britain has come from pure supplies in upland reservoirs or from underground sources and sometimes has been transferred considerable distances, for instance from Wales to Birmingham and from the Lake District to Manchester. These sources are rapidly proving inadequate and more water will in the future be extracted and stored from lowland rivers which are polluted by domestic, industrial, and agricultural effluents. These waters contain high concentrations of plant nutrients, which promote blooms of algae. In some nutrient-rich (eutrophic) reservoirs and lakes in North America and Europe, excessive plant production has lead to a change in fish populations probably because of oxygen lack during the decay of the algal blooms. Blue-green algae grow particularly well in eutrophic lakes and reservoirs. They accumulate in bays, where the effects of their decomposition are most severe, and also produce toxins which can kill fish and harm animals that drink the water.

Eutrophication could be controlled by reducing the concentrations of nutrients, but the nutrients that limit plant growth probably vary from area to area and their removal is frequently difficult or expensive. Phosphorus and nitrogen have been

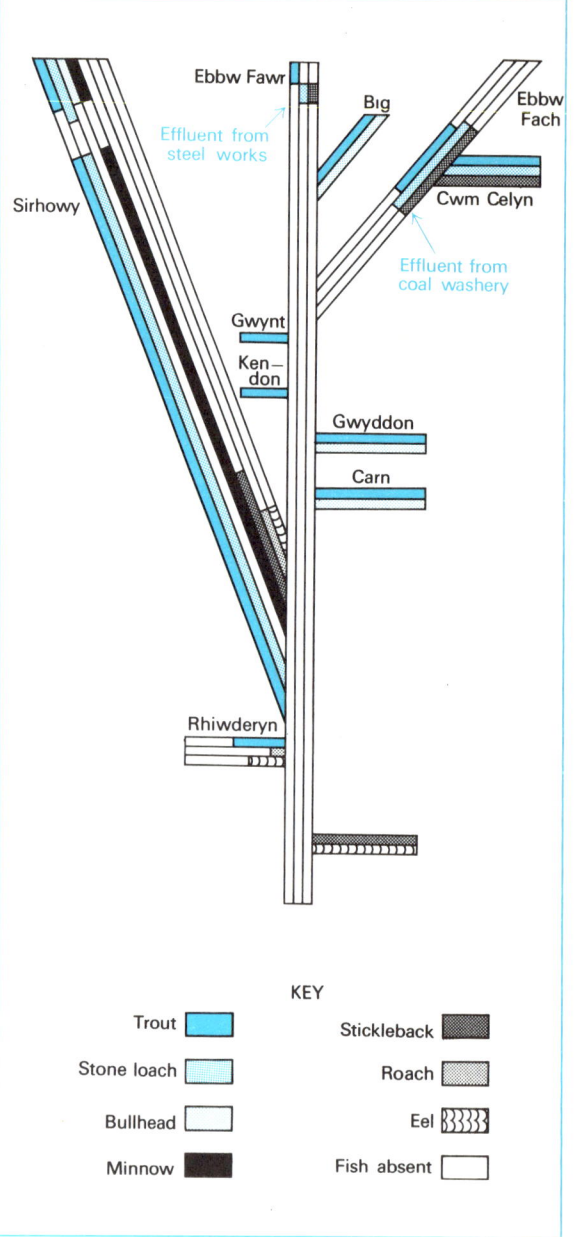

FIG. 9. Fish distribution in the River Ebbw and its tributaries. Although different fish species are found in different areas, and this is determined in part by the physical character of the stream, the Ebbw Fawr contains no fish downstream of the steel-works at Ebbw Vale and the Ebbw Fach contains no fish downstream of coal washeries at Abertillery. The absence of fish is due to pollution by industrial wastes including coal particles. (Data supplied by R. Williams and M. Harcup, U.W.I.S.T., Cardiff.)

12

implicated but other constituents, such as organic compounds, are probably equally important.

Phosphorus concentrations may be reduced in natural waters by removal at sewage-works using coagulants, or by reducing the use of phosphates in detergents, which generally contain 20–40 per cent tripolyphosphate. Fig. 8 shows the increase in concentration of phosphorus in the Thames and the Lea following the introduction of synthetic detergents.

Nitrogen is difficult to control as most of it comes from land drainage of fertilizers, and any reduction in the amount applied to the land would reduce agricultural production.

Algal growth is frequently limited by the poor supply of CO_2 which is renewed only slowly from the air or from carbonates. A bigger source of supply is bacterial respiration, and this is stimulated by high concentrations of organic matter such as those in sewage.

Eutrophication can be reduced by careful siting and design of drinking-water reservoirs. By building deep reservoirs the amount of light available per unit volume of water is reduced; by having different depths and points at which water can be drawn off, water can be removed from areas where algal concentrations are not high. Mixing the water in the reservoir by underwater jets prevents the development of anaerobic conditions in the bottom layer of water, and also reduces algal growth. Despite these safeguards it is sometimes necessary to use algicides, principally copper sulphate.

Effects on agriculture. Water for irrigation is generally taken from a nearby river and its bacterial and chemical quality is important. There has been recent concern about the increase in concentration of boron in some rivers containing sewage effluent. One important source of boron is as a bleach in detergent and many sewage effluents contain 3 mg/l of boron. Some crops like lettuce and many varieties of fruit are particularly sensitive to boron, requiring water containing less than 0·5 mg/l.

Effects on fish and fishing. In England and Wales about 5 per cent of rivers are fishless. The Ebbw, a river in South Wales polluted by steel-works, colliery, and sewage effluents, has no fish for about 90 per cent of its length (Fig. 9). Pollutants can be classified by their effects on fish. Some, like cyanide, ammonia, certain metals (copper, zinc, cadmium, nickel, lead), and pesticides, are directly toxic. Others, like oxidizable organic matter discharged in effluents, can result in the rapid growth of micro-

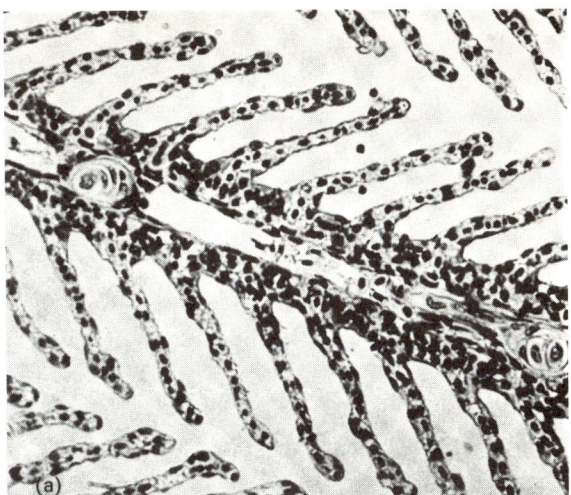

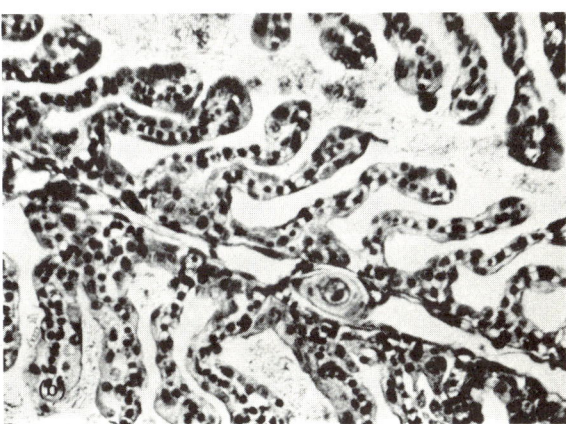

FIG. 10. Gills of trout living in (a) clean water, (b) water containing high concentrations of suspended matter (diatomaceous earth). Note thickening of epithelial cells and fusion of adjacent lamellae in (b). (From Herbert and Merkens (1961). *Int. J. Air Wat. Pollut.* 5.)

organisms and a consequent fall in the amount of oxygen dissolved in the water which may cause the death of fish. Suspended solids like coal washings, wood pulp, and china clay may exert their effect in several ways. In suspension they cause changes in the gill structure (Fig. 10) leading to the asphyxiation of fish; this rarely occurs at concentrations below 200 mg/l. When settled on the river bed the particles change the character of the fauna, causing a diminution in the food available to fish; and the interstices between stones become blocked and the spawning or development of species like bullheads, trout, and salmon is prevented.

13

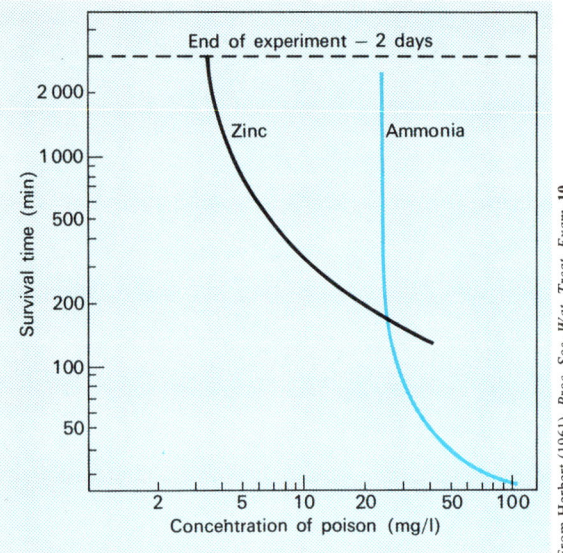

From Herbert (1961), *Proc. Soc. Wat. Treat. Exam.* **10**.

FIG. 11. Graph showing the survival time of trout in a range of concentrations of ammonia and zinc salts, in one set of experiments. As concentrations decrease the survival times increase. For convenience the scales in this diagram are logarithmic.

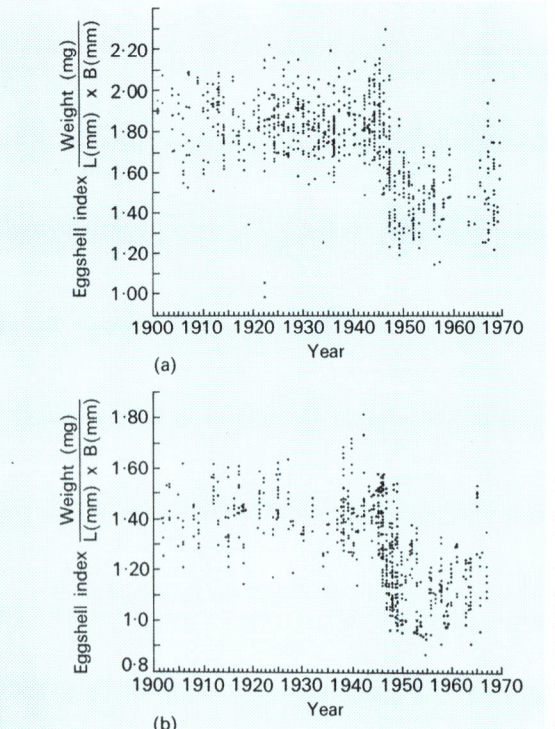

From Ratcliffe (1970), *J. appl. Ecol.* 7.

FIG. 12. Change in egg-shell index (an expression reflecting egg-shell thickness) of (a) peregrine falcon and (b) sparrow-hawk in Britain. Although highly variable, the egg-shell thickness clearly fell rapidly in both species between 1946 and 1948. The general introduction of DDT and gamma-BHC coincided with the onset of the egg-shell change.

These classes of pollutants are not distinct. The oxidation of organic matter produces ammonia as an intermediate product and in high enough concentrations this is toxic. Carbon dioxide is also formed and this can enhance the adverse effects of low oxygen concentrations. Pesticides like DDT which are toxic to fish over long periods of time at concentrations of 1 μg/l may have profound effects on other aquatic animals that influence fish indirectly by restricting their diet or by providing another and more concentrated source of pesticide.

Many laboratory experiments have been conducted to determine the concentrations of poisons at which fish die. Fig. 11 shows a typical result with ammonia and zinc. The concentration of poison is plotted against the time it takes for the 'average' fish to die. This kind of experiment helps form the basis of water quality standards for fisheries.

Some pollutants have adverse effects at concentrations far below that at which they are lethal. Some, like zinc, can be detected by fish at extremely low concentrations and avoided. Others, like Toxaphene, influence the ability of fish to learn. Copper inhibits the reproduction of at least one species (*Pimephales promelas*) at only 5 per cent of the lethal concentration and accelerates the development rate of embryos of rainbow trout.

Many environmental factors affect the toxicity of fish poisons. In hard waters at pH 8·5 only 9 mg/l of ammonia is required to kill trout whereas in water of a similar hardness at pH 6·5 at least 300 mg/l is needed. The explanation is that only the un-ionized molecule of ammonia is toxic and in acid waters most of the ammonia is in the ionized form. Copper and zinc are much less toxic in hard waters than in soft. Low temperature generally increases the toxicity of poisons, perhaps because the rate of entry is less affected by temperature than the rate at which poisons are excreted or detoxified in the body.

Pollutants are rarely present singly in natural waters and the toxicity of mixtures of poisons has to be assessed. Experiments have shown that poisons frequently act in an additive way. For example, 4 mg/l of phenol or 2 mg/l of zinc are toxic alone, but 2 mg/l of phenol and 1 mg/l of zinc are toxic when present together.